**First Step Math**

# Facts and Figures

For a free color catalog describing Gareth Stevens' list of high-quality books, call 1-800-542-2595 (USA) or 1-800-461-9120 (Canada). Gareth Stevens' Fax: (414) 225-0377.

**Library of Congress Cataloging-in-Publication Data**

Griffiths, Rose.
    Facts and figures/by Rose Griffiths; photographs by Peter Millard.
      p. cm. -- (First step math)
    Originally published: London: A&C Black, 1993, in series: Simple maths.
    Includes bibliographical references and index.
    ISBN 0-8368-1110-0
    1. Mathematical statistics--Juvenile literature.  2. Mensuration-
-Juvenile literature.  [1. Measurement.]  I. Millard, Peter, ill.
II. Title.  III. Series.
QA276.13.G75    1994
001.4'22--dc20                                                94-7984

This edition first published in 1994 by
**Gareth Stevens Publishing**
1555 North RiverCenter Drive, Suite 201
Milwaukee, Wisconsin 53212, USA

This edition © 1994 by Gareth Stevens, Inc. Original edition published in 1993 by A&C Black (Publishers) Ltd., 35 Bedford Row, London WC1R 4JH. © 1993 A&C Black (Publishers) Ltd. Additional end matter © 1994 by Gareth Stevens, Inc.

All rights to this edition reserved to Gareth Stevens, Inc. No part of this book may be reproduced, stored in a retrieval system, or transmitted in any form or by any means, electronic, mechanical, photocopying, recording, or otherwise, without the prior written permission of the publisher except for the inclusion of brief quotations in an acknowledged review.

Series editor: Patricia Lantier-Sampon
Editorial assistants: Mary Dykstra, Diane Laska
Mathematics consultant: Mike Spooner

Printed in the United States of America
1 2 3 4 5 6 7 8 9 99 98 97 96 95 94

First Step Math

# Facts and Figures

by Rose Griffiths
photographs by Peter Millard

Gareth Stevens Publishing
**MILWAUKEE**

These are our guinea pigs.

How are they different?

# How are they the same?

We had a vote to decide what to name our guinea pigs.

# Which names would you choose?

Boston and Nutmeg live in a hutch.

Every day we check to see if their hutch needs cleaning.

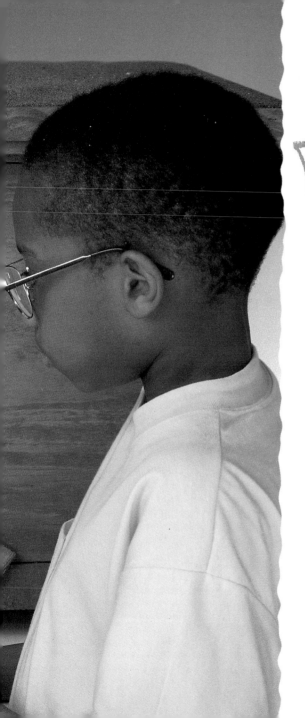

Whose turn is it to feed them?

| Monday | Jessica |
| Tuesday | Dean |
| Wednesday | Alice |
| Thursday | Georgia |
| Friday | David |
| Saturday | Owen |
| Sunday | Shaun |

Jessica has made a schedule.

## Carrot

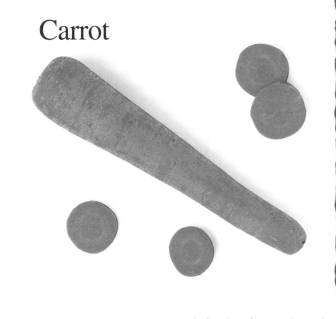

## Grass

## Which foods do our guinea pigs like best?

## Celery

## Turnip

We can't ask them, so we have to watch them.

Will Nutmeg choose carrots or grass to eat first?

We put a tally mark on our chart when Boston or Nutmeg eats something.

I'm counting in fives.

# Which food is their favorite?

# Which foods do you like?

**Menu**

Macaroni and Cheese
Stuffed peppers
Chop Suey
Lasagna
Salad
Falafel
Baked potatoes
Fried rice
Enchiladas

★

Fruit
Apple pie
Banana bread
Ice cream
Yogurt

Dean thinks Boston feels heavier than Nutmeg.

He could weigh them to find out.

When Nutmeg was a baby, she weighed about 3-1/2 ounces (100 grams).

Now she weighs a little more than 2 pounds (1 kilogram).

How much do you weigh?

Do you know what you weighed when you were born?

# We've made model guinea pigs . . .

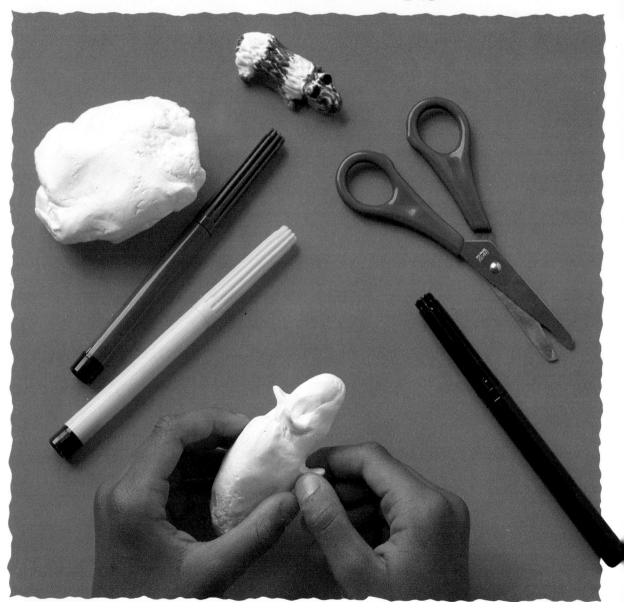

and toy hutches for them to live in.

# We've made pens for our guinea pigs, too.

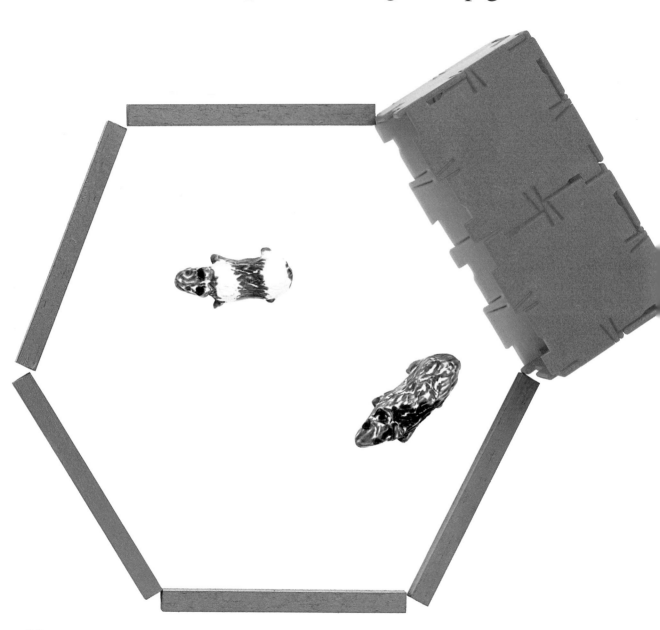

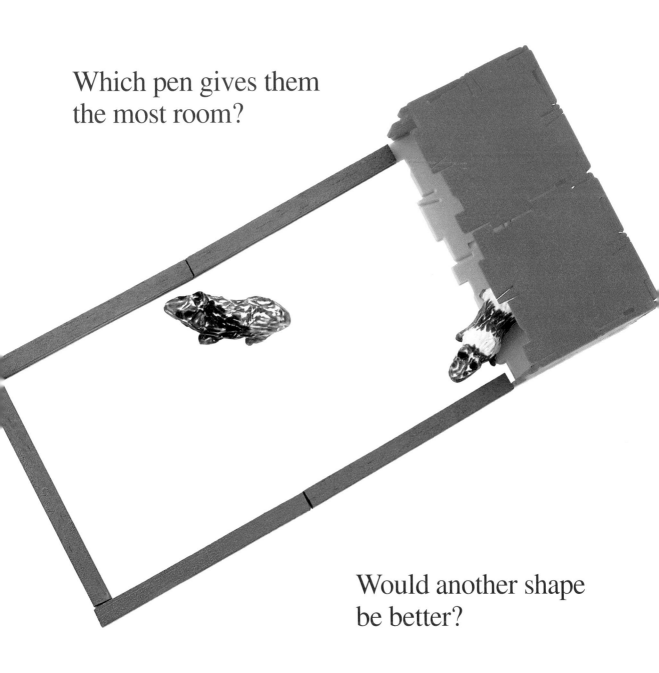

Which pen gives them the most room?

Would another shape be better?

We've made our own pet shop.

Do you have any pets?

What have you discovered about them?

# FOR MORE INFORMATION

## Notes for Parents and Teachers

**As you share this book with young readers, these notes may help you explain the mathematical principles behind the different activities.**

**pages 4, 5**
**Sorting and classifying**
Encourage children to point out similarities and differences they can see in the guinea pigs, such as color, length, and weight.

**pages 4, 5, 6, 7, 10, 11, 13, 14, 15, 23, 25  Surveys**
The information for a survey can be collected by observation or by direct questioning. Discuss with young readers ways of making a survey as fair and accurate as possible. For example, make sure each guinea pig is offered each of the four foods the same number of times. Before children begin to survey, they need to decide which questions to ask and which choices to offer.

**pages 9, 10, 11, 12**
**Ordering and combining**
Making a schedule gives children practice in putting things in order. On pages 11 and 12, the children have to decide how to give the guinea pigs two foods at a time, making sure they try out all four foods. There are six ways of combining two foods out of four — you can have the first food with any of three others, then the second with two others, and the third with one other (3+2+1=6). Figuring out all the possible combinations of events

is a preliminary step toward thinking about probability.

**page 12**
**Data collection sheets**
The example shown here uses tally marks grouped in fives. Children who are not yet confident about counting in fives may use single tally marks.

**pages 9, 13, 14**
**Representing data**
The bar chart on page 13 shows the frequency with which each food was chosen by the guinea pigs. The most frequent choice is called the mode. A menu is a useful source of information and an example of a simple database.

**pages 16, 17, 18, 19**
**Measuring weight**
A direct comparison of the weight of two guinea pigs can be made by holding them both or by carefully placing one on each pan of a balance. To record the weight of a growing animal from time to time, you need to use standard units of measurement, such as ounces or grams.

**pages 20, 21  Size and scale**
The model hutches are in the shape of a rectangular box and a triangular prism. Matching the size of the hutch to the size of the guinea pig encourages children to develop an understanding of scale and proportion.

**pages 22, 23  Shape and area**
The easiest way to compare the areas of two model pens is to draw them on paper, cut them out, and put one on top of the other. The hexagonal pen on page 22 has a larger area than the rectangular one on page 23.

## Things to Do

### 1. Favorite names
How was your name chosen for you? What is the most popular name among your friends or in your school? You can do a survey to find out.

### 2. Meal puzzle
Try this puzzle. Our guinea pigs like carrots, turnips, grass, dandelions, and celery. If we give them two foods out of these five for each meal, how many different meals can we give them? Make a list of as many meals as you can think of.

### 3. Pets and homes
Make models of pets such as cats, dogs, guinea pigs, or rabbits with self-hardening clay. Then make a basket or hutch for your pet to sleep in. How big does the hutch or basket need to be? What shape will you make it?

### 4. Pet shop visit
Visit a pet shop that has live animals with your parents or teacher and count the number of dogs. Then count the number of cats, birds, gerbils, and other types of animals. Put a mark on a chart you have made beforehand for each animal. Which pet is the most common? Which pet is the least common?

### 5. Plant facts and figures
Plant some bean seeds and keep a mathematical record of what happens. Write down what you observe each day. Be sure to measure your bean sprouts and stalks daily and write these exact measurements on your chart.

## Fun Facts about Facts and Figures

**1.** Mathematics is the study of numbers and the relationships that can exist between different numbers — such as size, shape, and order. Mathematics is based on logical reasoning and can be used in problem-solving.

**2.** Balances were used by the Egyptians to weigh objects accurately as early as 5000 B.C.

**3.** Guinea pigs are not really pigs — they are rodents. The capybara is the largest rodent and can weigh up to 250 pounds (113 kilograms). The smallest rodent is the tiny pygmy mouse, which weighs less than 1 ounce (28 grams).

**4.** Public opinion polls are surveys that measure what people think about various subjects. Each person's opinion counts as one vote, or tally. If polls are done fairly, the numbers from these polls can be helpful to companies, politicians, voters, and others who are interested in what the general public thinks about different issues.

**5.** There are many types of charts, such as tables, graphs, and diagrams. If these charts are set up fairly and accurately, they can provide a wide range of valuable information to students and working people.

**6.** Databases are stored collections of information that are organized especially for, and made available to, people who use computers.

# Glossary

**charts** — sheets or posters that display several pieces of information in a way that is easy to understand.

**choose** — to select or pick out.

**guinea pigs** — small rodents that are commonly kept as pets. Guinea pigs are related to the wild cavies of South America.

**heavy** — weighing a lot.

**hutch** — a home for a rabbit, guinea pig, or other small pet.

**models** — small copies of living things or inanimate objects, such as animals, boats, or cars.

**pens** — small, enclosed areas in which animals can live and play.

**schedule** — a list or plan for something that must be done and who is responsible for doing it.

**tally marks** — little lines or marks that stand for a number. Tally marks provide a fast, easy way to keep track of information.

**vote** — to choose or select someone or something as your personal favorite.

**weigh** — to measure how heavy something is.

## Places to Visit

Everything we do involves some basic mathematical principles. Listed below are a few museums that offer a wide variety of mathematical information and experiences. You may also be able to locate other museums in your area. Just remember: you don't always have to visit a museum to experience the wonders of mathematics. Math is everywhere!

The Smithsonian Institution
1000 Jefferson Drive SW
Washington, D.C. 20560

The Exploratorium
3601 Lyon Street
San Francisco, CA 94123

Royal British Columbia Museum
675 Belleville Street
Victoria, British Columbia
V8V 1X4

Ontario Science Center
770 Don Mills Road
Don Mills, Ontario
M3C 1T3

Museum of Science and Industry
57th Street and Lake Shore Drive
Chicago, IL 60637

## More Books to Read

*Anno's Math Games*
    Mitsumasa Anno
    (Philomel Books)

*Counting on Frank*
    Rod Clement
    (Gareth Stevens)

*How Many? How Much?*
  Monica Weiss
  (Troll Associates)

*In the Toy Store*
  Vincent O'Connor
  (Raintree)

*My First Number Book*
  Rod Clement
  (Gareth Stevens)

*Ten Beads Tall*
  M. Twinn
  (National Curriculum)

## Videotapes

*Learning About Numbers*
  (My Sesame Street Videos,
  Children's Television
  Workshop)

*What Is the Biggest
  Living Thing?*
  (Coronet)

## Index

charts 12
counting 12

feeding 9
foods 9, 10, 13, 14, 15

guinea pigs 4-5, 6-7, 8, 10-11, 16, 17, 18, 20, 22

hutches 8, 21

models 20, 21, 22, 23

naming 6, 7

pens 22, 23
pet shop 24

schedules 9

shapes 22, 23

tally marks 12

vote 6

weighing 16, 17, 18, 19

LEE COUNTY LIBRARY SYSTEM
DISCARD
3 3262 00150 9154

J001.4
G
Griffiths
FActs and figures

LEE COUNTY LIBRARY
107 Hawkins Ave
Sanford, N. C. 27330

GAYLORD S